Contents

Introduction

Indoor air pollution poses many challenges to the health professional. This booklet offers an overview of those challenges, focusing on acute conditions, with patterns that point to particular agents and suggestions for appropriate remedial action.

The individual presenting with environmentally associated symptoms is apt to have been exposed to airborne substances originating not outdoors, but indoors. Studies from the United States and Europe show that persons in industrialized nations spend more than 90 percent of their time indoors[1]. For infants, the elderly, persons with chronic diseases, and most urban residents of any age, the proportion is probably higher. In addition, the concentrations of many pollutants indoors exceed those outdoors. The locations of highest concern are those involving prolonged, continuing exposure—that is, the home, school, and workplace.

The lung is the most common site of injury by airborne pollutants. Acute effects, however, may also include non-respiratory signs and symptoms, which may depend upon toxicological characteristics of the substances and host-related factors.

Heavy industry-related occupational hazards are generally regulated and likely to be dealt with by an on-site or company physician or other health personnel[2]. This booklet addresses the indoor air pollution problems that may be caused by contaminants encountered in the daily lives of persons in their homes and offices. These are the problems more likely to be encountered by the primary health care provider.

Etiology can be difficult to establish because many signs and symptoms are nonspecific, making differential diagnosis a distinct challenge. Indeed, multiple pollutants may be involved. The challenge is further compounded by the similar manifestations of many of the pollutants and by the similarity of those effects, in turn, to those that may be associated with allergies, influenza, and the common cold. Many effects may also be associated, independently or in combination with, stress, work pressures, and seasonal discomforts.

Because a few prominent aspects of indoor air pollution, notably environmental tobacco smoke (p. 5) and "sick building syndrome" (p. 17), have been brought to public attention, individuals may volunteer suggestions of a connection between respiratory or other symptoms and conditions in the home or, especially, the workplace. Such suggestions should be seriously considered and pursued, with the caution that such attention could also lead to inaccurate attribution of effects. Questions listed in the diagnostic leads sections will help determine the cause of the health problem. The probability of an etiological association increases if the individual can convincingly relate the disappearance or lessening of symptoms to being away from the home or workplace.

How To Use This Booklet

The health professional should use this booklet as a tool in diagnosing an individual's signs and symptoms that could be related to an indoor air pollution problem. The document is organized according to pollutant or pollutant group. Key signs and symptoms from exposure to the pollutant(s) are listed, with diagnostic leads to help determine the cause of the health problem. A quick reference summary of this information is included in this booklet (pg.3). Remedial action is suggested, with comment providing more detailed information in each section. References for information included in each section are listed at the end of that section.

It must be noted that some of the signs and symptoms noted in the text may occur only in association with significant exposures, and that effects of lower exposures may be milder and more vague, unfortunately underscoring the diagnostic challenge. Further, signs and symptoms in infants and children may be atypical (some such departures have been specifically noted).

The reader is cautioned that this is not an all-inclusive reference, but a necessarily selective survey intended to suggest the scope of the problem. A detailed medical history is essential, and the diagnostic checklist (pg.4) may be helpful in this regard. Resolving the problem may sometimes require a multidisciplinary approach, enlisting the advice and assistance of others outside the medical profession. The references cited in the notes throughout and in the *For Assistance and Additional Information* section will provide the reader with additional information.

References

[1] U.S. Environmental Protection Agency, Office of Air and Radiation. Report to Congress on Indoor Air Quality, Volume II: Assessment and Control of Indoor Air Pollution, pp. i, 4-14. EPA-400-1-89-001C, 1989.

[2] The U.S. Environmental Protection Agency sets and enforces air quality standards only for ambient air. The Toxic Substances Control Act (TSCA) grants EPA broad authority to control chemical substances and mixtures that present an unreasonable risk of injury to health and environment. The Federal Insecticide, Fungicide, and Rodenticide Act (FIFRA) authorizes EPA to control pesticide exposures by requiring that any pesticide be registered with EPA before it may be sold, distributed, or used in this country. The Safe Drinking Water Act authorizes EPA to set and enforce standards for contaminants in

public water systems. EPA has set several standards for volatile organic compounds that can enter the air through volatilization from water used in a residence or other building. As to the indoor air in workplaces, two Federal agencies have defined roles concerning exposure to (usually single) substances. The National Institute for Occupational Safety and Health and Human Services (NIOSH), part of the Department of Health and Human Services, reviews scientific information, suggests exposure limitations, and recommends measures to protect workers' health. The Occupational Safety and Health Administration (OSHA), part of the Department of Labor, sets and enforces workplace standards. The U.S. Consumer Product Safety Commission (CPSC) regulates consumer products which may release indoor air pollutants. In the United States there are no Federal Standards that have been developed specifically for indoor air contaminants in non-occupational environments. There are, however, some source emission standards that specify maximum rates at which contaminants can be released from a source.

For more extensive information, see the publication cited above, in particular Chapter 7, "Existing Indoor Air Quality Standards", and Chapter 9, "Indoor Air Pollution Control Programs".

Diagnostic Quick Reference

Signs and Symptoms	Environmental Tobacco Smoke	Other Combustion Products	Biological Pollutants	Volatile Organics	Heavy Metals	Sick Bldg Syndrome
	pp.5	pp.7	pp.10	pp.13	pp.15	pp.17
Respiratory						
Rhinitis, nasal congestion	■	■	■	■		■
Epistaxis				■[1]		
Pharyngitis, cough	■	■	■	■		■
Wheezing, worsening asthma	■	■		■		■
Dyspnea	■[2]		■			■
Severe lung disease						■[3]
Other						
Conjunctival irritation	■	■	■	■		■
Headache or dizziness	■	■	■	■	■	■
Lethargy,fatigue, malaise		■[4]	■[5]	■	■	■
Nausea, vomiting, anorexia		■[4]	■	■	■	
Cognitive impairment, personality change		■[4]		■	■	■
Rashes			■	■	■	
Fever, chills			■[6]		■	
Tachycardia		■[4]			■	
Retinal hemorrhage		■[4]				
Myalgia				■[5]		■
Hearing loss				■		

[1]Associated especially with formaldehyde. [2]In asthma. [3]Hypersensitivity pneumonitis, Legionnaires' Disease. [4]Particularly associated with high CO levels. [5]Hypersensitivity pneumonitis, humidifier fever. [6]With marked hypersensitivity reactions and Legionnaires' Disease.

Particular Effects Seen in Infants and Children

Environmental Tobacco Smoke: frequent upper respiratory infections, otitis media; persistent middle-ear effusion; asthma onset, increased severity; recurrent pneumonia, bronchitis.

Acute Lead Toxicity: irritability, abdominal pain, ataxia, seizures, loss of consciousness.

Diagnostic Checklist

It is vital that the individual and the health care professional comprise a cooperative diagnostic team in analyzing diurnal and other patterns that may provide clues to a complaint's link with indoor air pollution. A diary or log of symptoms correlated with time and place may prove helpful. If an association between symptoms and events or conditions in the home or workplace is not volunteered by the individual, answers to the following questions may be useful, together with the medical history.

The health care professional can investigate further by matching the individual's signs and symptoms to those pollutants with which they may be associated, as detailed in the discussions of various pollutant categories.

- When did the [symptom or complaint] begin?
- Does the [symptom or complaint] exist all the time, or does it come and go? That is, is it associated with times of day, days of the week, or seasons of the year?
- (If so) Are you usually in a particular place at those times?
- Does the problem abate or cease, either immediately or gradually, when you leave there? Does it recur when you return?
- What is your work? Have you recently changed employers or assignments, or has your employer recently changed location?
- (If not) Has the place where you work been redecorated or refurnished, or have you recently started working with new or different materials or equipment? (These may include pesticides, cleaning products, craft supplies, et al.)
- What is the smoking policy at your workplace? Are you exposed to environmental tobacco smoke at work, school, home, etc.?
- Describe your work area.
- Have you recently changed your place of residence?
- (If not) Have you made any recent changes in, or additions to, your home?
- Have you, or has anyone else in your family, recently started a new hobby or other activity?
- Have you recently acquired a new pet?
- Does anyone else in your home have a similar problem? How about anyone with whom you work? (An affirmative reply may suggest either a common source or a communicable condition.)

NOTE: A more detailed exposure history form, developed by the U.S. Public Health Service's Agency for Toxic Substances and Disease Registry (ATSDR) in conjunction with the National Institute for Occupational Safety and Health, is available from: Allen Jansen, ATSDR, 1600 Clifton Road, N.E., Mail Drop E33, Atlanta, Georgia 30333, (404) 639-6205. Request "*Case Studies in Environmental Medicine #26: Taking an Exposure History.*" Continuing Medical Education Credit is available in conjunction with this monograph.

Health Problems Related To
Environmental Tobacco Smoke

Key Signs/Symptoms in Adults . . .

- rhinitis/pharyngitis, nasal congestion, persistent cough
- conjunctival irritation
- headache
- wheezing (bronchial constriction)
- exacerbation of chronic respiratory conditions

. . . and in Infants and Children

- asthma onset
- increased severity of, or difficulty in controlling, asthma
- frequent upper respiratory infections and/or episodes of otitis media
- persistent middle-ear effusion
- snoring
- repeated pneumonia, bronchitis

Diagnostic Leads

- Is individual exposed to environmental tobacco smoke on a regular basis?
- Test urine of infants and small children for cotinine, a biomarker for nicotine

Remedial Action

While improved general ventilation of indoor spaces may decrease the odor of environmental tobacco smoke (ETS), health risks cannot be eliminated by generally accepted ventilation methods. Research has led to the conclusion that total removal of tobacco smoke—a complex mixture of gaseous and particulate components—through general ventilation is not feasible.[3]

The most effective solution is to eliminate all smoking from the individual's environment, either through smoking prohibitions or by restricting smoking to properly designed smoking rooms. These rooms should be separately ventilated to the outside.[4]

Some higher efficiency air cleaning systems, under select conditions, can remove some tobacco smoke particles. Most air cleaners, including the popular desk-top models, however, cannot remove the gaseous pollutants from this source. And while some air cleaners are designed to remove specific gaseous pollutants, none is expected to remove all of them and should not be relied upon to do so. (For further comment, see p. 21.)

Comment

Environmental tobacco smoke is a major source of indoor air contaminants. The ubiquitous nature of ETS in indoor environments indicates that some unintentional inhalation of ETS by nonsmokers is unavoidable. Environmental tobacco smoke is a dynamic, complex mixture of more than 4,000 chemicals found in both vapor and particle phases. Many of these chemicals are known toxic or carcinogenic agents. Nonsmoker exposure to ETS-related toxic and carcinogenic substances will occur in indoor spaces where there is smoking.

All the compounds found in "mainstream" smoke, the smoke inhaled by the active smoker, are also found in "sidestream" smoke, the emission from the burning end of the cigarette, cigar, or pipe. ETS consists of both sidestream smoke and exhaled mainstream smoke. Inhalation of ETS is often termed "secondhand smoking", "passive smoking", or "involuntary smoking."

The role of exposure to tobacco smoke via active smoking as a cause of lung and other cancers, emphysema and other chronic obstructive pulmonary diseases, and cardiovascular and other diseases in adults has been firmly established[5,6,7]. Smokers, however, are not the only ones affected.

The U.S. Environmental Protection Agency (EPA) has classified ETS as a known human (Group A) carcinogen and estimates that it is responsible for approximately 3,000 lung cancer deaths per year among nonsmokers in the United States[8]. The U.S. Surgeon General, the National Research Council, and the National Institute for Occupational Safety and Health also concluded that passive smoking can cause lung cancer in otherwise healthy adults who never smoked[9,10,11].

Children's lungs are even more susceptible to harmful effects from ETS. In infants and young children up to three years, exposure to ETS causes an approximate doubling in the incidence of pneumonia, bronchitis, and bronchiolitis. There is also strong evidence of increased middle ear effusion, reduced lung function, and reduced lung growth. Several recent studies link ETS with increased incidence and prevalence of asthma and increased severity of asthmatic symptoms in children of mothers who smoke heavily. These respiratory illnesses in childhood may very well contribute to the small but significant lung function reductions associated with exposure to ETS in adults. The adverse health effects of ETS, especially in children, correlate with the amount of smoking in the home and are often more prevalent when both parents smoke[12].

The connection of children's symptoms with ETS may not be immediately evident to the clinician and may become apparent only after careful questioning. Measurement of biochemical markers such as cotinine (a metabolic nicotine derivative) in body fluids (ordinarily urine) can provide evidence of a child's exposure to ETS[13].

The impact of maternal smoking on fetal development has also been well documented. Maternal smoking is also associated with increased incidence of Sudden Infant Death Syndrome, although it has not been determined to what extent this increase is due to in utero versus postnatal (lactational and ETS) exposure[14].

Airborne particulate matter contained in ETS has been associated with impaired breathing, lung diseases, aggravation of existing respiratory and cardiovascular disease, changes to the body's immune system, and lowered defenses against inhaled particles[15]. For direct ETS exposure, measurable annoyance, irritation, and adverse health effects have been demonstrated in nonsmokers, children and spouses in particular, who spend significant time in the presence of smokers[16,17]. Acute cardiovascular effects of ETS include increased heart rate, blood pressure, blood carboxyhemoglobin; and related reduction in exercise capacity in those with stable angina and in healthy people. Studies have also found increased incidence of nonfatal heart disease among nonsmokers exposed to ETS, and it is thought likely that ETS increases the risk of peripheral vascular disease, as well[18].

References

[3] Leaderer, B.P., Cain, W.S., Isseroff, R., Berglund, L.G. "Ventilation Requirements in Buildings II". *Atmos. Environ.* 18:99-106.

See also: Repace, J.L. and Lowrey, A.H. "An indoor air quality standard for ambient tobacco smoke based on carcinogenic risk." *New York State Journal of Medicine* 1985; 85:381-83.

[4] American Society of Heating, Refrigeration and Air Conditioning Engineers. *Ventilation for Acceptable Air Quality*; ASHRAE Standard 62-1989.

[5] International Agency for Research on Cancer. *IARC Monographs On the Evaluation of the Carcinogenic Risk of Chemicals to Man, Vol. 38: Tobacco Smoking.* World Health Organization, 1986.

[6] U.S. Department of Health and Human Services. *Reducing the Health Consequences of Smoking: 25 Years of Progress, A Report of the Surgeon General.* DHHS Publication No. (CDC) 89-8411. 1989.

[7] U.S. Department of Health and Human Services. *The Health Benefits of Smoking Cessation, A Report of the Surgeon General.* DHHS Publication No. (CDC) 90-8416. 1990.

[8] U.S. Environmental Protection Agency, Office of Air and Radiation and Office of Research and Development. *Respiratory Health Effects of Passive Smoking: Lung Cancer and Other Disorders.* EPA 600-6-90-006F. 1992.

[9] U.S. Department of Health and Human Services. *The Health Consequences of Involuntary Smoking, A Report of the Surgeon General.* DHHS Publication No. (PHS) 87-8398. 1986.

[10] National Research Council, *Environmental Tobacco Smoke: Measuring Exposures and Assessing Health Effects.* National Academy Press. 1986.

[11] National Institute for Occupational Safety and Health. *Environmental Tobacco Smoke in the Workplace: Lung Cancer and Other Health Effects.* U.S. Department of Health and Human Services, Current Intelligence Bulletin 54. 1991.

[12] U.S. Environmental Protection Agency. *Respiratory Health Effects of Passive Smoking: Lung Cancer and Other Disorders.*

[13] U.S. Environmental Protection Agency. *Respiratory Health Effects of Passive Smoking: Lung Cancer and Other Disorders.*

[14] U.S. Environmental Protection Agency. *Respiratory Health Effects of Passive Smoking: Lung Cancer and Other Disorders.*

[15] Pope, C.A. III, Schwartz, J. and Ransom, M.R. "Daily Mortality and PM 10 Pollution in Utah, Salt Lake, and Cache Valleys". *Archives of Environmental Health* 1992: 46:90-96.

[16] U.S. Department of Health and Human Services. *The Health Consequences of Involuntary Smoking, A Report of the Surgeon General.*

[17] National Research Council. *Environmental Tobacco Smoke: Measuring Exposures and Assessing Health Effects.*

[18] American Heart Association Council on Cardiopulmonary and Critical Care. "Environmental Tobacco Smoke and Cardiovascular Disease." *Circulation* 1992; 86:1-4.

Health Problems Caused By

Other Combustion Products

(Stoves, Space Heaters, Furnaces, Fireplaces)

Key Signs/Symptoms

- dizziness or headache
- confusion
- nausea/emesis
- fatigue
- tachycardia
- eye and upper respiratory tract irritation
- wheezing/bronchial constriction
- persistent cough
- elevated blood carboxyhemoglobin levels
- increased frequency of angina in persons with coronary heart disease

Diagnostic Leads

- What types of combustion equipment are present, including gas furnaces or water heaters, stoves, unvented gas or kerosene space heaters, clothes dryers, fireplaces? Are vented appliances properly vented to the outside?
- Are household members exhibiting influenza-like symptoms during the heating season? Are they complaining of nausea, watery eyes, coughing, headaches?
- Is a gas oven or range used as a home heating source?
- Is the individual aware of odor when a heat source is in use?
- Is heating equipment in disrepair or misused? When was it last professionally inspected?
- Does structure have an attached or underground garage where motor vehicles may idle?
- Is charcoal being burned indoors in a hibachi, grill, or fireplace?

Remedial Action

Periodic professional inspection and maintenance of installed equipment such as furnaces, water heaters, and clothes dryers are recommended. Such equipment should be vented directly to the outdoors. Fireplace and wood or coal stove flues should be regularly cleaned and inspected before each heating season. Kitchen exhaust fans should be exhausted to outside. Vented appliances should be used whenever possible. Charcoal should never be burned inside. Individuals potentially exposed to combustion sources should consider installing carbon monoxide detectors that meet the requirements of Underwriters Laboratory (UL) Standard 2034. No detector is 100% reliable, and some individuals may experience health problems at levels of cardon monoxide below the detection sensitivity of these devices.

Comment

Aside from environmental tobacco smoke, the major combustion pollutants that may be present at harmful levels in the home or workplace stem chiefly from malfunctioning heating devices, or inappropriate, inefficient use of such devices. Incidents are largely seasonal. Another source may be motor vehicle emissions due, for example, to proximity to a garage (or a loading dock located near air intake vents).

A variety of particulates, acting as additional irritants or, in some cases, carcinogens, may also be released in the course of combustion. Although faulty venting in office buildings and other nonresidential structures has resulted in combustion product problems, most cases involve the home or non-work-related consumer activity. Among possible sources of contaminants: gas ranges that are malfunctioning or used as heat sources; improperly flued or vented fireplaces, furnaces, wood or coal stoves, gas water heaters and gas clothes dryers; and unvented or otherwise improperly used kerosene or gas space heaters.

The gaseous pollutants from combustion sources include some identified as prominent atmospheric pollutants—carbon monoxide (CO), nitrogen dioxide (NO_2), and sulfur dioxide (SO_2).

Carbon monoxide is an asphyxiant. An accumulation of this odorless, colorless gas may result in a varied constellation of symptoms deriving from the compound's affinity for and combination with hemoglobin, forming carboxyhemoglobin (COHb) and disrupting oxygen transport. The elderly, the fetus, and persons with cardiovascular and pulmonary diseases are particularly sensitive to elevated CO levels. Methylene chloride, found in some common household products, such as paint strippers, can be metabolized to form carbon monoxide which combines with hemoglobin to form COHb. The following chart shows the relationship between CO concentrations and COHb levels in blood.

Tissues with the highest oxygen needs—myocardium, brain, and exercising muscle—are the first affected. Symptoms may mimic influenza and include fatigue, headache, dizziness,

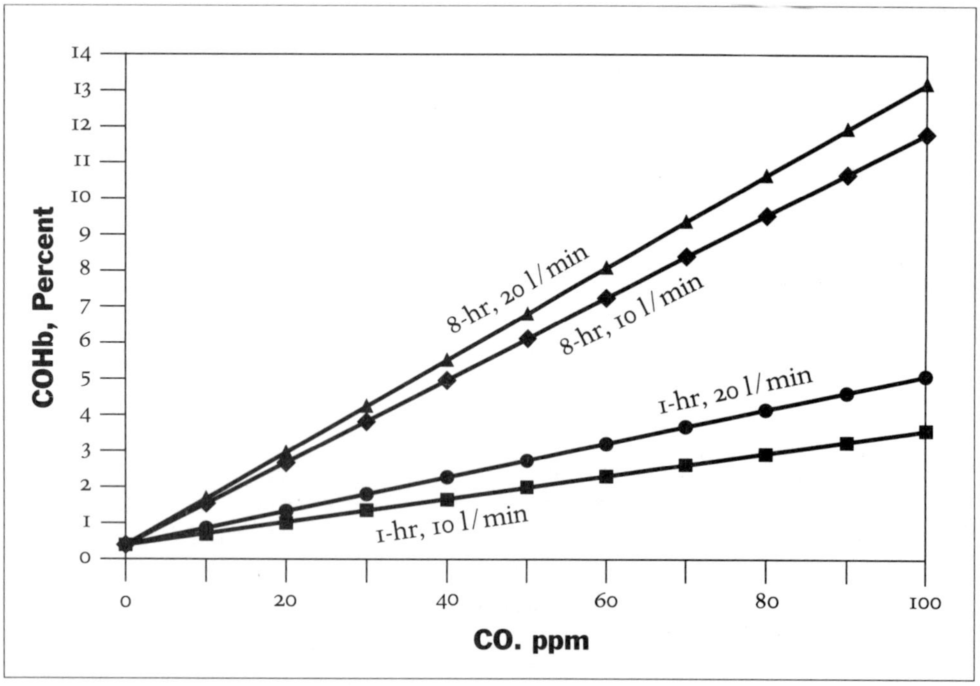

Relationship between carbon monoxide (CO) concentrations and carboxy-hemoglobin (COHb) levels in blood

Predicted COHb levels resulting from 1- and 8-hr exposures to carbon monoxide at rest (10 l/min) and with light exercise (20 l/min) are based on the Coburn-Foster-Kane equation using the following assumed parameters for nonsmoking adults: altitude = 0 ft; initial COHb level = 0.5%; Haldane constant = 218; blood volume = 5.5 l; hemoglobin level = 15 g/100 ml; lung diffusivity = 30 ml/torr/min; endogenous rate = 0.007 ml/min.

Source: Raub, J.A. and Grant, L.D. 1989. "Critical health issues associated with review of the scientific criteria for carbon monoxide." Presented at the 82nd Annual Meeting of the Air Waste Management Association. June 25–30. Anaheim, CA. Paper No. 89.54.1. Used with permission.

nausea and vomiting, cognitive impairment, and tachycardia. Retinal hemorrhage on funduscopic examination is an important diagnostic sign[19], but COHb must be present before this finding can be made, and the diagnosis is not exclusive. Studies involving controlled exposure have also shown that CO exposure shortens time to the onset of angina in exercising individuals with ischemic heart disease and decreases exercise tolerance in those with chronic obstructive pulmonary disease (COPD)[20].

Note: Since CO poisoning can mimic influenza, the health care provider should be suspicious when an entire family exhibits such symptoms at the start of the heating season and symptoms persist with medical treatment and time.

Nitrogen dioxide and *sulfur dioxide* act mainly as irritants, affecting the mucosa of the eyes, nose, throat, and respiratory tract. Acute SO_2-related bronchial constriction may also occur in people with asthma or as a hypersensitivity reaction. Extremely high-dose exposure (as in a building fire) to NO_2 may result in pulmonary edema and diffuse lung injury. Continued exposure to high NO_2 levels can contribute to the development of acute or chronic bronchitis.

The relatively low water solubility of NO_2 results in minimal mucous membrane irritation of the upper airway. The principal site of toxicity is the lower respiratory tract. Recent studies indicate that low-level NO_2 exposure may cause

Carboxyhemoglobin levels and related health effects.

% COHb in blood	Effects associated with this COHb level
80	Death[a]
60	Loss of consciousness; death if exposure continues[a]
40	Confusion; collapse on exercise[a]
30	Headache; fatigue; impaired judgement[a]
7–20	Statistically significant decreased maximal oxygen consumption during strenuous exercise in healthy young men[b]
5–17	Statistically significant diminution of visual perception, manual dexterity, ability to learn, or performance in complex sensorimotor tasks (such as driving)[b]
5–5.5	Statistically significant decreased maximal oxygen consumption and exercise time during strenuous exercise in young healthy men[b]
Below 5	No statistically significant vigilance decrements after exposure to CO[b]
2.9–4.5	Statistically significant decreased exercise capacity (i.e., shortened duration of exercise before onset of pain) in patients with angina pectoris and increased duration of angina attacks[b]
2.3–4.3	Statistically significant decreased (about 3–7%) work time to exhaustion in exercising healthy men[b]

Source: [a] U.S. EPA (1979); [b] U.S. EPA (1985)

increased bronchial reactivity in some asthmatics, decreased lung function in patients with chronic obstructive pulmonary disease, and an increased risk of respiratory infections, especially in young children.

The high water solubility of SO_2 causes it to be extremely irritating to the eyes and upper respiratory tract. Concentrations above six parts per million produce mucous membrane irritation. Epidemiologic studies indicate that chronic exposure to SO_2 is associated with increased respiratory symptoms and decrements in pulmonary function[21]. Clinical studies have found that some asthmatics respond with bronchoconstriction to even brief exposure to SO_2 levels as low as 0.4 parts per million[22].

References

[19] Samet, J.M., Marbury, Marian C., and Spengler, J.D. "Health Effects and Sources of Indoor Air Pollution, Part I." *American Review of Respiratory Disease* 1987; 136:1486-1508.

[20] American Thoracic Society. "Report of the ATS Workshop on Environmental Controls and Lung Disease, Santa Fe, New Mexico, March 24-26, 1988." *American Review of Respiratory Disease* 1990; 142:915-39.

[21] Lipsett, M. "Oxides of Nitrogen and Sulfur." *Hazardous Materials Technology* 1992; 000:964-69.

[22] U.S. Environmental Protection Agency. "Review of the National Ambient Air Quality Standards for Sulfur Oxides: Updated Assessment of Scientific and Technical Information; Supplement to the 1986 Staff Paper Addendum (July 1993)."

Health Problems Caused By

Animal Dander, Molds, Dust Mites, Other Biologicals

Key Signs/Symptoms

- recognized infectious disease
- exacerbation of asthma
- rhinitis
- conjunctival inflammation
- recurrent fever
- malaise
- dyspnea
- chest tightness
- cough

Diagnostic Leads

Infectious disease:

- Is the case related to the workplace, home, or other location? (Note: It is difficult to associate a single case of any infectious disease with a specific site of exposure.)
- Does the location have a reservoir or disseminator of biologicals that may logically lead to exposure?

Hypersensitivity disease:

- Is the relative humidity in the home or workplace consistently above 50 percent?
- Are humidifiers or other water-spray systems in use? How often are they cleaned? Are they cleaned appropriately?
- Has there been flooding or leaks?
- Is there evidence of mold growth (visible growth or odors)?
- Are organic materials handled in the workplace?
- Is carpet installed on unventilated concrete (e.g., slab on grade) floors?
- Are there pets in the home?
- Are there problems with cockroaches or rodents?

Toxicosis and/or irritation:

- Is adequate outdoor air being provided?
- Is the relative humidity in the home or workplace above 50 percent or below 30 percent?
- Are humidifiers or other water-spray systems in use?
- Is there evidence of mold growth (visible growth or odors)?
- Are bacterial odors present (fishy or locker-room smells)?

Remedial Action

Provide adequate outdoor air ventilation to dilute human source aerosols.

Keep equipment water reservoirs clean and potable water systems adequately chlorinated, according to manufacturer instructions. Be sure there is no standing water in air conditioners. Maintain humidifiers and dehumidifiers according to manufacturers' instructions.

Repair leaks and seepage. Thoroughly clean and dry water-damaged carpets and building materials within 24 hours of damage, or consider removal and replacement.

Keep relative humidity below 50 percent. Use exhaust fans in bathrooms and kitchens, and vent clothes dryers to outside.

Control exposure to pets.

Vacuum carpets and upholstered furniture regularly. *Note:* While it is important to keep an area as dust-free as possible, cleaning activities often re-suspend fine particles during and immediately after the activity. Sensitive individuals should be cautioned to avoid such exposure, and have others perform the vacuuming, or use a commercially available HEPA (High Efficiency Particulate Air) filtered vacuum.

Cover mattresses. Wash bedding and soft toys frequently in water at a temperature above 130°F to kill dust mites.

Comment

Biological air pollutants are found to some degree in every home, school, and workplace. Sources include outdoor air and human occupants who shed viruses and bacteria, animal occupants (insects and other arthropods, mammals) that shed allergens, and indoor surfaces and water reservoirs where fungi and bacteria can grow, such as humidifiers[23]. A number of factors allow biological agents to grow and be released into the air. Especially important is high relative humidity, which encourages house dust mite populations to increase and allows fungal growth on damp surfaces. Mite and fungus contamination can be caused by flooding, continually damp carpet (which may occur when carpet is installed on poorly ventilated concrete floors), inadequate exhaust of bathrooms, or kitchen-generated moisture[24]. Appliances such as humidifiers, dehumidifiers, air conditioners, and drip pans under cooling coils (as in refrigerators), support the growth of bacteria and fungi.

Components of mechanical heating, ventilating, and air conditioning (HVAC) systems may also serve as reservoirs or sites of microbial amplification[25]. These include air intakes near potential sources of contamination such as standing water, organic debris or bird droppings, or integral parts of the mechanical system itself, such as various humidification systems, cooling coils, or condensate drain pans. Dust and debris may be deposited in the duct work or mixing boxes of the air handler.

Biological agents in indoor air are known to cause three types of human disease: infections, where pathogens invade human tissues, hypersensitivity diseases, where specific activation of the immune system causes disease; and toxicoses, where biologically produced chemical toxins cause direct toxic effects.

In addition, exposure to conditions conducive to biological contamination (e.g., dampness, water damage) has been related to nonspecific upper and lower respiratory symptoms. Evidence is available that shows that some episodes of the group of nonspecific symptoms known as "sick building syndrome" (see p. 17) may be related to microbial contamination in buildings[26].

Tuberculosis

The transmission of airborne infectious diseases is increased where there is poor indoor air quality[27,28]. The rising incidence of tuberculosis is at least in part a problem associated with crowding and inadequate ventilation. Evidence is increasing that inadequate or inappropriately designed ventilation systems in health care settings or other crowded conditions with high-risk populations can increase the risk of exposure[29].

The incidence of tuberculosis began to rise in the mid-1980s, after a steady decline. The 1989 increase of 4.7 percent to a total of 23,495 cases in the United States was the largest since national reporting of the disease began in 1953, and the number of cases has continued to increase each year[30]. Fresh air ventilation is an important factor in contagion control. Such procedures as sputum induction and collection, bronchoscopy, and aerosolized pentamidine treatments in persons who may be at risk for tuberculosis (e.g., AIDS patients) should be carried out in negative air pressure areas, with air exhausted directly to the outside and away from intake sources[31]. Unfortunately, many health care facilities are not so equipped. Properly installed and maintained ultraviolet irradiation, particularly of upper air levels in an indoor area, is also a useful means of disinfection[32].

Legionnaires' Disease

A disease associated with indoor air contamination is Legionnaires' Disease, a pneumonia that primarily attacks exposed people over 50 years old, especially those who are immunosuppressed, smoke, or abuse alcohol. Exposure to especially virulent strains can also cause the disease in other susceptible populations. The case fatality rate is high, ranging from five to 25 percent. Erythromycin is the most effective treatment. The agent, Legionella pneumophila, has been found in association with cooling systems, whirlpool baths, humidifiers, food market vegetable misters, and other sources, including residential tap water[33]. This bacterium or a closely related strain also causes a self-limited (two- to five-day), flu-like illness without pneumonia, sometimes called Pontiac Fever, after a 1968 outbreak in that Michigan city.

Allergic Reactions

A major concern associated with exposure to biological pollutants is allergic reactions, which range from rhinitis, nasal congestion, conjunctival inflammation, and urticaria to asthma. Notable triggers for these diseases are allergens derived from house dust mites; other arthropods, including cockroaches; pets (cats, dogs, birds, rodents); molds; and protein-containing furnishings, including feathers, kapok, etc. In occupational settings, more unusual allergens (e.g., bacterial enzymes, algae) have caused asthma epidemics. Probably most proteins of non-human origin can cause asthma in a subset of any appropriately exposed population[34].

The role of mites as a source of house dust allergens has been known for 20 years[34,35]. It is now possible to measure mite allergens in the environment and IgE antibody levels in patients using readily available techniques and standardized protocols. Experts have proposed provisional standards for levels of mite allergens in dust that lead to sensitization and symptoms. A risk level where chronic exposure may cause sensitization is 2μg Der pI (Dermatophagoides pteronysinus allergen I) per gram of dust (or 100 mites/g or 0.6 mg quanine/g of dust). A risk level for acute asthma in mite-allergic individuals is 10μg (Der pI) of the allergen per gram of dust (or 500 mites/g of dust).

Controlling house dust mite infestation includes covering mattresses, hot washing of bedding, and removing carpet from bedrooms. For mite allergic individuals, it is recommended that home relative humidities be lower than 45 percent. Mites desiccate in drier air (absolute humidities below 7 kg.). Vacuum cleaning and use of acaricides can be effective short-term remedial strategies. One such acaracide, Acarosan, is registered with EPA to treat carpets, furniture, and beds for dust mites.

Hypersensitivity Pneumonitis

Another class of hypersensitivity disease is hypersensitivity pneumonitis, which may include humidifier fever. Hypersensitivity pneumonitis, also called allergic alveo-litis, is a granulomatous interstitial lung disease caused by exposure to airborne antigens. It may affect from one to five percent or more of a specialized population exposed to appropriate antigens (e.g., farmers and farmers' lung, pigeon breeders and pigeon breeders' disease)[37]. Continued antigen exposure may lead to end-stage pulmonary fibrosis. Hypersensitivity pneumonitis is frequently misdiagnosed as a pneumonia of infectious etiology. The prevalence of hypersensitivity pneumonitis in the general population is unknown.

Outbreaks of hypersensitivity pneumonitis in office buildings have been traced to air conditioning and humidification systems contaminated with bacteria and molds[38]. In the home, hypersensitivity pneumonitis is often caused by contaminated humidifiers or by pigeon or pet bird antigens. The period of sensitization before a reaction occurs may be as long as months or even years. Acute symptoms, which occur four to six hours postexposure and recur on challenge with the offending agent, include cough, dyspnea, chills, myalgia, fatigue, and high fever. Nodules and nonspecific infiltrates may be noted on chest films. The white blood cell count is elevated, as is specific IgG to the offending antigen. Hypersensitivity pneumonitis generally responds to corticosteroids or cessation of exposure (either keeping symptomatic people out of contaminated environments or removing the offending agents).

Humidifier Fever

Humidifier fever is a disease of uncertain etiology[39]. It shares symptoms with hypersensitivity pneumonitis, but the high attack rate and short-term effects may indicate that toxins (e.g., bacterial endotoxins) are involved. Onset occurs a few hours after exposure. It is a flu-like illness marked by fever, headache, chills, myalgia, and malaise but without prominent pulmonary symptoms. It normally subsides within 24 hours without residual effects, and a physician is rarely consulted. Humidifier fever has been related to exposure to amoebae, bacteria, and fungi found in humidifier reservoirs, air conditioners, and aquaria. The attack rate within a workplace may be quite high, sometimes exceeding 25 percent.

Bacterial and fungal organisms can be emitted from impeller (cool mist) and ultrasonic humidifiers. Mesophilic fungi, thermophilic bacteria, and thermophilic actinomycetes—all of which are associated with development of allergic responses—have been isolated from humidifiers built into the forced-air heating system as well as separate console units. Airborne concentrations of microorganisms are noted during operation and might be quite high for individuals using ultrasonic or cool mist units. Drying and chemical disinfection with bleach or 3% hydrogen peroxide solution are effective remedial measures over a short period, but cannot be considered as reliable maintenance. Only rigorous, daily, and end-of-season cleaning regimens, coupled with disinfection, have been shown to be effective. Manual cleaning of contaminated reservoirs can cause exposure to allergens and pathogens.

Mycotoxins

Another class of agents that may cause disease related to indoor airborne exposure is the mycotoxins. These agents are fungal metabolites that have toxic effects ranging from short-term irritation to immunosuppression and cancer. Virtually all the information related to diseases caused by mycotoxins concerns ingestion of contaminated food[40]. However, mycotoxins are contained in some kinds of fungus spores, and these can enter the body through the respiratory tract. At least one case of neurotoxic symptoms possibly related to airborne mycotoxin exposure in a heavily contaminated environment has been reported[41]. Skin is another potential route of exposure to mycotoxins. Toxins of several fungi have caused cases of severe dermatosis. In view of the serious nature of the toxic effects reported for mycotoxins, exposure to mycotoxin-producing agents should be minimized.

References

[23] Burge, Harriet A. and Feely, J.C. "Indoor Air Pollution and Infectious Diseases." In: Samet, J.M. and Spengler, J.D. eds., *Indoor Air Pollution, A Health Perspective* (Baltimore MD: Johns Hopkins University Press, 1991), pp. 273-84.

[24] Brunekreeff, B., Dockery, D.W. et al. "Home Dampness and Respiratory Morbidity in Children." *American Review of Respiratory Disease* 1989; 140:1363-67.

[25] Berstein, R.S., Sorenson, W.G. et al. "Exposures to Respirable Airborne Penicillium from a Contaminated Ventilation System: Clinical, Environmental, and Epidemiological Aspects." *American Industrial Hygiene Association Journal* 1983; 44:161-69.

[26] Burge, Harriet A. "Bioaerosols: Prevalence and Health Effects in the Indoor Environment." *Journal of Allergy and Clinical Immunology* 1990; 86:687-704.

[27] Burge, Harriet A. "Risks Associated With Indoor Infectious Aerosols." *Toxicology and Industrial Health* 1990; 6:263-73.

[28] Brundage, J.F., Scott, R. et al. "Building-Associated Risk of Febrile Acute Respiratory Disease in Army Trainees." *Journal of the American Medical Association* 1988; 259:2108-12.

[29] Nolan, C.M., Elarth, A.M. et al. "An Outbreak of Tuberculosis in a Shelter for Homeless Men: A Description of Its Evolution and Control." *American Review of Respiratory Disease* 1991; 143:257-61.

[30] American Lung Association. *Lung Disease Data 1993*. Publication No. 0456, 1993.

[31] Centers for Disease Control and American Thoracic Society. *Core Curriculum On Tuberculosis*. Second Edition, 1991.

[32] Nardell, E.A., Keegan, Joann et al. "Airborne Infection: Theoretical Limits of Protection Achievable By Building Ventilation." *American Review of Respiratory Disease* 1991; 144:302-06.

[33] Lee, T.C., Stout, Janet E. and Yu, V.L. "Factors Predisposing to Legionella pneumophila Colonization in Residential Water Systems." *Archives of Environmental Health* 1988; 43:59-62.

[34] Weissman, D.N. and Schuyler, M.R. "Biological Agents and Allergic Diseases." In: Samet, J.M. and Spengler, J.D. eds., *Indoor Air Pollution, A Health Perspective* (Baltimore MD: Johns Hopkins University Press, 1991), pp. 285-305.

[35] Arlian, L.G. "Biology and Ecology of House Dust Mite, Dermatophagoldes spp. and Euroglyphus spp." *Immunology and Allergy Clinics of North America* 1989; 9:339-56.

[36] Platts-Mills, T.A.E. and Chapman, M.D. "Dust Mites: Immunology, Allergic Disease, and Environmental Control." *Journal of Allergy and Clinical Immunology* 1987; 80:755-75.

[37] Fink, J.N."Hypersensitivity Pneumonitis." In: Middleton, E., Reed, C.E. and Ellis, E.F. eds., *Allergy Principles and Practice* (St. Louis: C.V. Mosby, 19xx), pp. 1085-1100.

[38] Fink, J.N. "Hypersensitivity Pneumonitis." In: Middleton, E., Reed, C.E. and Ellis, E.F. eds., *Allergy Principles and Practice* (St. Louis: C.V. Mosby, 19xx), pp. 1085-1100.

[39] Burge, Harriet A., Solomon, W.R. and Boise, J.R. "Microbial Prevalence in Domestic Humidifiers." *Applied and Environmental Microbiology* 1980; 39:840-44.

[40] Baxter, C.S., Wey, H.E. and Burg, W.R. "A Prospective Analysis of the Potential Risk Associated with Inhalation of Aflatoxin-Contaminated Grain Dusts." *Food and Cosmetics Toxicology* 1981; 19:763-69.

[41] Croft, W.A., Jarvia, B.B. Yatawara, C.S. 1986. Airborne outbreak of trichothecene toxicosis. *Atmosph. Environ.* 20:549-552. See also Baxter, C.S. Wey, H.E., Burg, W.E. 1981. A prospective analysis of the potential risk associated with inhalation of aflatoxin-contaminated grain dusts. Food Cosmet Toxicol. 19:763-769.

Health Problems Caused By

Volatile Organic Compounds

(Formaldehyde, Pesticides, Solvents, Cleaning Agents)

Key Signs/Symptoms

- conjunctival irritation
- nose, throat discomfort
- headache
- allergic skin reaction
- dyspnea
- declines in serum cholinesterase levels
- nausea, emesis
- epistaxis (formaldehyde)
- fatigue
- dizziness

Diagnostic Leads

- Does the individual reside in mobile home or new conventional home containing large amounts of pressed wood products?
- Has individual recently acquired new pressed wood furniture?
- Does the individual's job or avocational pursuit include clerical, craft, graphics, or photographic materials?
- Are chemical cleaners used extensively in the home, school, or workplace?
- Has remodeling recently been done in home, school or workplace?
- Has individual recently used pesticides, paints, or solvents?

Remedial Action

Increase ventilation when using products that emit volatile organic compounds, and meet or exceed any label precautions. Do not store opened containers of unused paints and similar materials within home or office. See special note on pesticides.

Formaldehyde is one of the best known volatile organic compound (VOC) pollutants, and is one of the few indoor air pollutants that can be readily measured. Identify, and if possible, remove the source if formaldehyde is the potential cause of the problem. If not possible, reduce exposure: use polyurethane or other sealants on cabinets, paneling and other furnishings. To be effective, any such coating must cover all surfaces and edges and remain intact. Formaldehyde is also used in permanent press fabric and mattress ticking. Sensitive individuals may choose to avoid these products.

Comment

At room temperature, volatile organic compounds are emitted as gases from certain solids or liquids. VOCs include a variety of chemicals (e.g., formaldehyde, benzene, perchloroethylene), some of which may have short- and long-term effects. Concentrations of many VOCs are consistently higher indoors than outdoors. A study by the EPA, covering six communities in various parts of the United States, found indoor levels up to ten times higher than those outdoors—even in locations with significant outdoor air pollution sources, such as petrochemical plants[42].

A wide array of volatile organics are emitted by products used in home, office, school, and arts/crafts and hobby activities. These products, which number in the thousands, include:

- personal items such as scents and hair sprays;
- household products such as finishes, rug and oven cleaners, paints and lacquers (and their thinners), paint strippers, pesticides (see below);
- dry-cleaning fluids
- building materials and home furnishings;
- office equipment such as some copiers and printers;
- office products such as correction fluids and carbonless copy paper[43,44];
- graphics and craft materials including glues and adhesives, permanent markers, and photographic solutions.

Many of these items carry precautionary labels specifying risks and procedures for safe use; some do not. Signs and symptoms of VOC exposure may include eye and upper respiratory irritation, rhinitis, nasal congestion, rash, pruritus, headache, nausea, vomiting, dyspnea and, in the case of formaldehyde vapor, epistaxis.

Formaldehyde

Formaldehyde has been classified as a probable human carcinogen by the EPA[45]. Urea-formaldehyde foam insulation (UFFI), one source of formaldehyde used in home construction until the early 1980s, is now seldom installed, but formaldehyde-based resins are components of finishes, plywood, paneling, fiberboard, and particleboard, all widely employed in mobile and conventional home construction as building materials (subflooring, paneling) and as components of furniture and cabinets, permanent press fabric, draperies, and mattress ticking.

Airborne formaldehyde acts as an irritant to the conjunctiva and upper and lower respiratory tract. Symptoms are temporary and, depending upon the level and length of exposure, may range from burning or tingling sensations in eyes, nose, and throat to chest tightness and wheezing. Acute, severe reactions to formaldehyde vapor—which has a distinctive, pungent odor—may be associated with hypersensitivity. It is estimated that 10 to 20 percent of the U.S. population, including asthmatics, may have hyperreactive airways which may make them more susceptible to formaldehyde's effects[46].

Pesticides

Pesticides sold for household use, notably impregnated strips, and foggers or "bombs", which are technically classed as semivolatile organic compounds, include a variety of chemicals in various forms. Exposure to pesticides may cause harm if they are used improperly. However, exposure to pesticides via inhalation of spray mists may occur during normal use. Exposure can also occur via inhalation of vapors and contaminated dusts after use (particularly to children who may be in close contact with contaminated surfaces). Symptoms may include headache, dizziness, muscular weakness, and nausea. In addition, some pesticide active ingredients and inert components are considered possible human carcinogens. Label directions must be explicitly followed[47].

References

[42] U.S. Environmental Protection Agency, Office of Acid Deposition, Environmental Monitoring and Quality Assurance. *Project Summary: The Total Exposure Assessment Methodology (TEAM) Study.* EPA-600-S6-87-002, 1987.
[43] Marks, J.G., Jr. Trautlein, J.J. et al. "Contact Urticaria and Airway Obstruction From Carbonless Copy Paper." *Journal of the American Medical Association* 1984; 252:1038-40.
[44] LaMarte, F.P., Merchant, J.A. and Casale, T.B. "Acute Systemic Reactions to Carbonless Copy Paper Associated With Histamine Release." *Journal of the American Medical Association* 1988; 260:242-43.
[45] U.S. Environmental Protection Agency, Office of Air and Radiation. *Report to Congress on Indoor Air Quality, Volume II: Assessment and Control of Indoor Air Pollution*, pp. i, 4-14. EPA-400-1-89-001C, 1989.
[46] U.S. Environmental Protection Agency, U.S. Public Health Service, and National Environmental Health Association. *Introduction to Indoor Air Quality: A Reference Manual*, p. 87. EPA-400-3-91-003, 1991.
[47] U.S. Environmental Protection Agency Office of Research and Development. *Final Report: Nonoccupational Pesticide Exposure Study (NOPES)*, p. 60. EPA-600-3-90-003, 1990.

Health Problems Caused By
Heavy Metals: Airborne Lead and Mercury Vapor

Key Signs/Symptoms of Lead Poisoning in Adults . . .

- gastrointestinal discomfort/constipation/anorexia/nausea
- fatigue, weakness
- personality changes
- headache
- hearing loss
- tremor, lack of coordination

. . . and in Infants and Small Children

- irritability
- abdominal pain
- ataxia
- seizures/loss of consciousness
- (chronic) learning deficits
- hyperactivity, reduced attention span

Key Signs/Symptoms of Mercury Poisoning

- muscle cramps or tremors
- headache
- tachycardia
- intermittent fever
- acrodynia
- personality change
- neurological dysfunction

Diagnostic Leads

- Does the family reside in old or restored housing?
- Has renovation work been conducted in the home, workplace, school, or day care facility?
- Is the home located near a busy highway or industrial area?
- Does the individual work with lead materials such as solder or automobile radiators?
- Does the child have sibling, friend, or classmate recently diagnosed with lead poisoning?
- Has the individual engaged in art, craft, or workshop pursuits?
- Does the individual regularly handle firearms?
- Has the home interior recently been painted with latex paint that may contain mercury?
- Does the individual use mercury in religious or cultural activities?

Remedial Action

Wet-mop and wipe furniture frequently to control lead dust. Have professional remove or encapsulate lead containing paint; individuals involved in this and other high exposure activities should use appropriate protective gear and work in well-ventilated areas. Do not burn painted or treated wood.

Comment

Airborne Lead

Most health professionals are aware of the threat of lead toxicity, particularly its long term impact on children in the form of cognitive and developmental deficits which are often cumulative and subtle. Such deficits may persist into adulthood[48]. According to the American Academy of Pediatrics, an estimated three to four million children in the U.S. under age six have blood lead levels that could cause impaired development, and an additional 400,000 fetuses are at similar risk[49].

Lead toxicity may alternatively present as acute illness. Signs and symptoms in children may include irritability, abdominal pain, emesis, marked ataxia, and seizures or loss of consciousness. In adults, diffuse complaints—including headache, nausea, anorexia (and weight loss), constipation, fatigue, personality changes, and hearing loss—coupled with exposure opportunity may lead to suspicion of lead poisoning.

Lead inhibits heme synthesis. Since interruption of that process produces protoporphyrin accumulation at the cellular level, the standard screening method is investigation of blood lead (PbB) levels which reveal recent exposure to lead. Acute symptomology in adults is often associated with PbB at levels of 40 μg/dl or higher. There is good evidence for adverse effects of lead in very young children at much lower levels.[50,51] The Centers for Disease Control and Prevention has set 10 μg/dl as the level of concern[52]. Increased maternal Pb exposure has also been deemed significant in pregnancy, since an umbilical cord PbB of greater than 10 μg/dl has been correlated with early developmental deficits. If sufficiently high PbB levels are confirmed, chelation therapy may be indicated. Suspected low level lead contamination cannot be accurately identified by a erythrocyte protoporphyrin (EP) finger-stick test, but requires blood lead analysis.

Lead poisoning via ingestion has been most widely publicized, stressing the roles played by nibbling of flaking paint by infants and toddlers and by the use of lead-containing foodware (glass, and soldered metal-ceramic ware) by adults. Lead dust flaking or "chalking" off lead painted walls generated by friction surfaces is a major concern. Airborne lead, however, is also a worrisome source of toxicity. There is no skin absorption associated with inorganic lead.

Airborne lead outdoors, originating chiefly from gasoline additives, has been effectively controlled since the 1980s through regulation at the federal level. Much of this lead still remains in the soil near heavily trafficked highways and in urban areas, however, and can become airborne at times. It may enter dwellings via windows and doors, and contaminated soil can also be tracked inside.

Indoors, the chief source is paint. Lead levels in paints for interior use have been increasingly restricted since the 1950s, and many paints are now virtually lead free. But older housing and furniture may still be coated with leaded paint, sometimes surfacing only after layers of later, non-lead paint have flaked away or have been stripped away in the course of restoration or renovation. In these circumstances, lead dust and fumes can permeate the air breathed by both adults and children.

Additional sources of airborne lead include art and craft materials, from which lead is not banned, but the U.S. Consumer Product Safety Commission (CPSC) requires its presence to be declared on the product label if it is present in toxic amounts. Significant quantities are found in many paints and glazes, stained glass, as well as in some solder. Hazardous levels of atmospheric lead have been found at police and civilian firing ranges. Repair and cleaning of automobile radiators in inadequately ventilated premises can expose workers to perilous levels of airborne lead. The use of treated or painted wood in fireplaces or improperly vented wood stoves may release a variety of substances, including lead and other heavy metals, into the air.

Mercury Vapor

While old paint has been the most publicized source of airborne heavy metal (i.e., lead), new paint has emerged as a concern as well. A 1990 report detailed elevated levels of mercury in persons exposed to interior latex (water-based) paint containing phenylmercuric acetate[53]. PMA was a preservative that was used to prolong the product's shelf life.

Initial action by the U.S. Environmental Protection Agency resulted in the elimination of mercury compounds from indoor latex paints at the point of manufacture as of August 1990, with the requirement that paints containing mercury, including existing stocks originally designed for indoor use, be labeled or relabeled "For Exterior Use Only". As of September 1991, phenylmercuric acetate is forbidden in the manufacture of exterior latex paints as well. Latex paints containing hazardous levels of mercury may still remain on store shelves or in homes where they were left over after initial use, however.

An additional matter of concern, recently noted by the CPSC, is the sprinkling of mercury about the home by some ethnic/religious groups[54]. According to the CPSC, mercury for this purpose is purveyed by some herbal medicine or botanical shops to consumers unaware of the dangers of the substance.

References

[48] Needleman, H.L. Schell, A. et al. "The Long-Term Effects of Exposure to Low Doses of Lead in Childhood: An 11-Year Follow-up Report." *The New England Journal of Medicine* 1990; 322:83-88.
[49] American Academy of Pediatrics. "Lead Poisoning: Next Focus of Environmental Action." Statement issued January 1991.
[50] Bellinger, D., Sloman, J. et al. "Low-Level Lead Exposure and Children's Cognitive Function in the Preschool Years." *Pediatrics* 1991; 87:219-27.
[51] "Lower 'Threshold of Concern' for Children's Lead Levels". *FDA Consumer*, December 1991. p. 6.
[52] Centers for Disease Control. "Preventing Lead Poisoning in Young Children". October 1991.
[53] Agocs, Mary M., Etzel, Ruth A. et al. "Mercury Exposure from Latex Interior Paint." *The New England Journal of Medicine*. 1990; 323:1096-1101.
[54] Consumer Product Safety Commission. Safety Alert: Mercury Vapors.

Health Problems Caused By

Sick Building Syndrome

Key Signs/Symptoms

- lethargy or fatigue
- headache, dizziness, nausea
- irritation of mucous membranes
- sensitivity to odors

Diagnostic Leads

- Are problems temporally related to time spent in a particular building or part of a building?
- Do symptoms resolve when the individual is not in the building?
- Do symptoms recur seasonally (heating, cooling)?
- Have co-workers, peers noted similar complaints?

Remedial Action

Appropriate persons—employer, building owner or manager, building investigation specialist, if necessary state and local government agency medical epidemiologists and other public health officials—should undertake investigation and analysis of the implicated building, particularly the design and operation of HVAC systems, and correct contributing conditions. Persistence on the part of individual(s) and health care consultant(s) may be required to diagnose and remediate the building problems.

Comment

The term "sick building syndrome" (SBS), first employed in the 1970s, describes a situation in which reported symptoms among a population of building occupants can be temporally associated with their presence in that building. Typically, though not always, the structure is an office building.

Generally, a spectrum of specific and nonspecific complaints are involved. Typical complaints, in addition to the signs and symptoms already listed, may also include eye and/or nasopharyngeal irritation, rhinitis or nasal congestion, inability to concentrate, and general malaise—complaints suggestive of a host of common ailments, some ubiquitous and easily communicable. The key factors are commonality of symptoms and absence of symptoms among building occupants when the individuals are not in the building.

Sick building syndrome should be suspected when a substantial proportion of those spending extended time in a building (as in daily employment) report or experience acute on-site discomfort. It is important, however, to distinguish SBS from problems of building related illness. The latter term is reserved for situations in which signs and symptoms of diagnosable illness are identified and can be attributed directly to specific airborne building contaminants. Legionnaires' Disease and hypersensitivity pneumonitis, for example, are building related illnesses.

There has been extensive speculation about the cause or causes of SBS. Poor design, maintenance, and/or operation of the structure's ventilation system may be at fault[55]. The ventilation system itself can be a source of irritants. Interior redesign, such as the rearrangement of offices or installation of partitions, may also interfere with efficient functioning of such systems.

Another theory suggests that very low levels of specific pollutants, including some discussed in the preceding pages, may be present and may act synergistically, or at least in combination, to cause health effects. Humidity may also be a factor: While high relative humidity may contribute to biological pollutant problems, an unusually low level—below 20 or 30 percent—may heighten the effects of mucosal irritants and may even prove irritating itself. Other contributing elements may include poor lighting and adverse ergonomic conditions, temperature extremes, noise, and psychological stresses that may have both individual and interpersonal impact.

The prevalence of the problem is unknown. A 1984 World Health Organization report suggested that as many as 30 percent of new and remodeled buildings worldwide may generate excessive complaints related to indoor air quality[56]. In a nationwide, random sampling of U.S. office workers, 24 percent perceived air quality problems in their work environments, and 20 percent believed their work performance was hampered thereby[57].

When SBS is suspected, the individual physician or other health care provider may need to join forces with others (e.g., clinicians consulted by an individual's co-workers, as well as industrial hygienists and public health officials) to adequately investigate the problem and develop appropriate solutions.

References

[55] A professional group, the American Society of Heating, Refrigerating, and Air-Conditioning Engineers (ASHRAE), has established standards of ventilation for the achievement of acceptable indoor air quality. These criteria do not have the force of law, are typically invoked only for new or renovated construction, and even when met do not assure comfortable and healthful air quality under all conditions and in all circumstances.

[56] U.S. Environmental Protection Agency, Office of Air and Radiation. *Indoor Air Facts No. 4: Sick Building Syndrome,* revised, 1991.

[57] Kreiss, Kathleen. "The Sick Building Syndrome: Where Is the Epidemiologic Basis?" *American Journal of Public Health* 1990; 80:1172-73.

Health Problems Caused By

Two Long-Term Risks: Asbestos and Radon

Asbestos and radon are among the most publicized indoor air pollutants. Both are known human carcinogens. Their carcinogenic effects are not immediate but are evident only years, even decades, after prolonged exposure.

Asbestos

Once widely used in structural fireproofing, asbestos may be found predominantly in heating systems and acoustic insulation, in floor and ceiling tiles, and in shingles in many older houses. It was formerly used in such consumer products as fireplace gloves, ironing board covers, and certain hair dryers.

When asbestos-containing material is damaged or disintegrates with age, microscopic fibers may be dispersed into the air. Over as long as twenty, thirty, or more years, the presence of these fibers within the lungs may result in asbestosis (asbestos-caused fibrosis of the lung, seen as a result of heavy occupational exposure)[58], lung cancer and pleural or peritoneal cancer, or mesothelioma[59]. For lung cancer, the effect of tobacco smoking in combination with asbestos exposure appears to be synergistic by approximately fivefold[60]. Occupational exposure may also be associated with increased risk of gastrointestinal malignancies. Attention should be focused on those populations with continual exposure and documented health effects, e.g. maintenance workers.

Products and materials containing asbestos are not necessarily so labeled. Construction professionals or state or local environmental agencies may inspect and analyze suspect materials. Manufacturers of particular products may also be able to supply information.

The risk of disease depends on exposure to airborne asbestos fibers. Average levels in buildings are low, and the risk to building occupants is therefore low.

Removal of asbestos is not always the best choice to reduce exposure. The EPA requires asbestos removal only in order to prevent significant public exposure and generally recommends an in-place management program when asbestos has been discovered and is in good condition[61].

Radon

Radon is the second leading cause of lung cancer, following smoking. Radon is odorless, colorless, and tasteless. It is a naturally occurring radioactive gas resulting from the decay of radium, itself a decay product of uranium. Radon in turn breaks down into radon decay products, short-lived radionuclides. These decay products, either free or attached to airborne particles, are inhaled, and further decay can take place in the lungs before removal by clearance mechanisms.

It is the emission of high-energy alpha particles during the radon decay process that increases the risk of lung cancer. While the risk to underground miners has long been known, the potential danger of residential radon pollution has been widely recognized only since the late 1970s, with the documentation of high indoor levels.

When radon decay products are inhaled and deposited in the lungs, the alpha emissions penetrate the cells of the epithelium lining the lung. Energy deposited in these cells during irradiation is believed to initiate the process of carcinogenesis. The EPA, the National Cancer Institute, the Centers for Disease Control and Prevention, and others estimate that thousands of lung cancer deaths per year are attributable to radon, based on data from epidemiologic studies of thousands of underground miners and from animal studies. Lung cancer is presently the only commonly accepted disease risk associated with radon.

Tobacco smoke in combination with radon exposure has a synergistic effect. Smokers and former smokers are believed to be at especially high risk. Scientists estimate that the increased risk of lung cancer to smokers from radon exposure is ten to twenty times higher than to people who have never smoked.

The EPA estimates that as many as six million homes throughout the country have elevated levels of radon. Since 1988, EPA and the Office of the Surgeon General have recommended that homes below the third floor be tested for radon.

Short term testing is the quickest way to determine if a potential problem exists, taking from two to ninety days to complete. Low-cost radon test kits are available by mail order, in hardware stores, and through other retail outlets[62].

Measurement devices should be state-certified or display the phrase, "Meets EPA Requirements". Trained contractors who meet EPA's requirements can also provide testing services. The most commonly used devices are charcoal canisters, electret ion detectors, alpha track detectors, and continuous monitors placed by contractors. Short term testing should be conducted in the lowest lived in area of the home, with the doors and windows shut. Long term testing can take up to a full year but is more likely to reflect the home's year round average radon level than short term testing. Alpha track detectors and electret ion detectors are the most common long-term testing devices.

Corrective steps include sealing foundation cracks and holes, and venting radon-laden air from beneath the foundation. Professional expertise should be sought for effective execution of these measures.

References

[58] The first death attributed to occupational asbestos exposure occurred in 1924; the details were recently recounted: Selikoff, I.J. and Greenberg, M. "A Landmark Case in Asbestosis." *Journal of the American Medical Association* 1991; 265:898-901.

[59] For a detailed discussion of asbestos-related pulmonary disease, see: Rom, W.N., Travis, W.D. and Brody, A.R. "Cellular and Molecular Basis of the Asbestos-related Diseases." *American Review of Respiratory Disease* 1991; 143:408-22.

[60] U.S. Environmental Protection Agency, Office of Research and Development. Airborne Asbestos Health Assessment Update. EPA-600-8-84-003F. June 1986.

[61] "Asbestos in Your Home", American Lung Association, U.S. Consumer Product Safety Commission, U.S. Environmental Protection Agency. September 1990. ALA Publication No. 3716

[62] See Samet, J.M., Marbury, Marian C. and Spengler, J.D. "Health Effects and Sources of Indoor Air Pollution, Part II." *American Review of Respiratory Disease* 1988; 137:221-42. This continuation of the overview cited earlier provides a table of commercial sources of testing equipment for sampling and monitoring levels of a variety of indoor air pollutants, including radon.

Questions That May Be Asked

The subject of indoor air pollution is not without some controversy. Indoor air quality is an evolving issue; it is important to keep informed about continuing developments in this area. The following questions may be asked of physicians and other health professionals.

What is "multiple chemical sensitivity" or "total allergy"?
The diagnostic label of multiple chemical sensitivity (MCS)—also referred to as "chemical hypersensitivity" or "environmental illness"— is being applied increasingly, although definition of the phenomenon is elusive and its pathogenesis as a distinct entity is not confirmed. Multiple chemical sensitivity has become more widely known and increasingly controversial as more patients received the label[63].

Persons with the diagnostic label of multiple chemical sensitivity are said to suffer multi-system illness as a result of contact with, or proximity to, a spectrum of substances, including airborne agents. These may include both recognized pollutants discussed earlier (such as tobacco smoke, formaldehyde, et al.) and other pollutants ordinarily considered innocuous. Some who espouse the concept of MCS believe that it may explain such chronic conditions as some forms of arthritis and colitis, in addition to generally recognized types of hypersensitivity reactions.

Some practitioners believe that the condition has a purely psychological basis. One study[63] reported a 65 percent incidence of current or past clinical depression, anxiety disorders, or somatoform disorders in subjects with this diagnosis compared with 28 percent in controls. Others, however, counter that the disorder itself may cause such problems[64], since those affected are no longer able to lead a normal life, or that these conditions stem from effects on the nervous system[65].

The current consensus is that in cases of claimed or suspected MCS, complaints should not be dismissed as psychogenic, and a thorough workup is essential. Primary care givers should determine that the individual does not have an underlying physiological problem and should consider the value of consultation with allergists and other specialists.

Who are "clinical ecologists"?
"Clinical ecology", while not a recognized conventional medical specialty, has drawn the attention of health care professionals as well as laypersons. The organization of clinical ecologists—physicians who treat individuals believed to be suffering from "total allergy" or "multiple chemical sensitivity"—was founded as the Society for Clinical Ecology and is now known as the American Academy of Environmental Medicine. Its ranks have attracted allergists and physicians from other traditional medical specialties[66].

What are ionizers and other ozone generating air cleaners?
Ion generators act by charging the particles in a room so that they are attracted to walls, floors, tabletops, draperies, occupants, etc. Abrasion can result in these particles being resuspended into the air. In some cases these devices contain a collector to attract the charged particles back to the unit. While ion generators may remove small particles (e.g., those in tobacco smoke) from the indoor air, they do not remove gases or odors, and may be relatively ineffective in removing large particles such as pollen and house dust allergens. Although some have suggested that these devices provide a benefit by rectifying a hypothesized ion imbalance, no controlled studies have confirmed this effect.

Ozone, a lung irritant, is produced indirectly by ion generators and some other electronic air cleaners and directly by ozone generators. While indirect ozone production is of concern, there is even greater concern with the direct, and purposeful introduction of a lung irritant into indoor air. There is no difference, despite some marketers' claims, between ozone in smog outdoors and ozone produced by these devices. Under certain use conditions ion generators and other ozone generating air cleaners can produce levels of this lung irritant significantly above levels thought harmful to human health. A small percentage of air cleaners that claim a health benefit may be regulated by FDA as a medical device. The Food and Drug Administration has set a limit of 0.05 parts per million of ozone for medical devices. Although ozone can be useful in reducing odors and pollutants in unoccupied spaces (such as removing smoke odors from homes involved in fires) the levels needed to achieve this are above those generally thought to be safe for humans.

Can other air cleaners help?
Ion generators and ozone generators are types of air cleaners; others include mechanical filter air cleaners, electronic air cleaners (e.g., electrostatic precipitators), and hybrid air cleaners utilizing two or more techniques. Generally speaking, existing air

cleaners are not appropriate single solutions to indoor air quality problems, but can be useful as an adjunct to effective source control and adequate ventilation. Air cleaning alone cannot adequately remove all pollutants typically found in indoor air.

The value of any air cleaner depends upon a number of factors, including its basic efficiency, proper selection for the type of pollutant to be removed, proper installation in relation to the space, and faithful maintenance. Drawbacks, varying with type, may include inadequate pollutant removal, re-dispersement of pollutants, deceptive masking rather than removal, generation of ozone, and unacceptable noise levels.

The EPA and CPSC have not taken a position either for or against the use of these devices in the home[67].

Should I have my ducts cleaned?

As awareness of the importance of indoor air quality grows, more people are looking at duct cleaning as a way to solve indoor air quality problems. Individuals considering having ducts cleaned should determine that contaminated ducts are the cause of their health problems. Even when contaminants are found in ducts, the source may lie elsewhere, and cleaning ducts may not permanently solve the problem. The duct cleaning industry is expanding to meet demand, using extensive advertising to encourage people to use their services. Individuals who employ such services should verify that the service provider takes steps to protect individuals from exposure to dislodged pollutants and chemicals used during the cleaning process. Such steps may range from using HEPA filtration on cleaning equipment, providing respirators for workers, and occupants vacating the premises during cleaning.

Can carpet make people sick?

Like many other household products and furnishings, new carpet can be a source of chemical emissions. Carpet emits volatile organic compounds, as do products that accompany carpet installation such as adhesives and padding. Some people report symptoms such as eye, nose and throat irritation; headaches; skin irritations; shortness of breath or cough; and fatigue, which they may associate with new carpet installation. Carpet can also act as a "sink" for chemical and biological pollutants including pesticides, dust mites, and fungi.

Individuals purchasing new carpet should ask retailers for information to help them select lower emitting carpet, cushion, and adhesives. Before new carpet is installed, they should ask the retailer to unroll and air out the carpet in a clean, well-ventilated area. They should consider leaving the premises during and immediately after carpet installation or schedule the installation when the space is unoccupied. Opening doors and windows and increasing the amount of fresh air indoors will reduce exposure to most chemicals released from newly installed carpet. During and after installation in a home, use of window fans and room air conditioners to exhaust fumes to the outdoors is recommended. Ventilation systems should be in proper working order, and should be operated during installation, and for 48 to 72 hours after the new carpet is installed.

Individuals should request that the installer follow the Carpet and Rug Institute's installation guidelines[68]. If new carpet has an objectionable odor, they should contact their carpet retailer. Finally, carpet owners should follow the manufacturer's instructions for proper carpet maintenance.

Can plants control indoor air pollution?

Recent reports in the media and promotions by the decorative houseplant industry characterize plants as "nature's clean air machine", claiming that National Aeronautics and Space Administration (NASA) research shows plants remove indoor air pollutants. While it is true that plants remove carbon dioxide from the air, and the ability of plants to remove certain other pollutants from water is the basis for some pollution control methods, the ability of plants to control indoor air pollution is less well established. Most research to date used small chambers without any air exchange which makes extrapolation to real world environments extremely uncertain. The only available study of the use of plants to control indoor air pollutants in an actual building could not determine any benefit from the use of plants[69]. As a practical means of pollution control, the plant removal mechanisms appear to be inconsequential compared to common ventilation and air exchange rates. In other words, the ability of plants to actually improve indoor air quality is limited in comparison with provision of adequate ventilation.

While decorative foliage plants may be aesthetically pleasing, it should be noted that overdamp planter soil conditions may actually promote growth of unhealthy micro-organisms.

References

[63] Black, D.W. Rathe, Ann and Goldstein, Rise B. "Environmental Illness: A Controlled Study of 26 Subjects With '20th Century Disease'." *Journal of the American Medical Association* 1990; 264:3166-70.

[64] Fiedler, N., Maccia, C., Kipen, H. "Evaluation of Chemically Sensitive Patients". *Journal of Occupational Medicine.* 1992. 34:529-538

[65] Heuser, G., Wojdani, A., Heuser, S. "Diagnostic Markers of Multiple Chemical Sensitivity". *Multiple Chemical Sensitivities: Addendum to Biologic Markers in Immunotoxicology.* 1992. pp. 117-138. National Research Council. National Academy Press. Washington D.C.

[66] See Ducataman et al. "What is Environmental Medicine?" *Journal of Occupational Medicine* 1990; 32: 1130-32. Also see American College of Physicians Health and Public Policy Committee. "Occupational and Environmental Medicine: The Internist's Role". Annals of Internal Medicine 1990; 113:974-82.

[67] For further specifics, see: U.S. Environmental Protection Agency, Office of Air and Radiation. *Residential Air Cleaning Devices — A Summary of Available Information.* EPA-400-1-90-002, 1990.

[68] Residential Carpet Installation Standard. The Carpet and Rug Institute. First Edition. 1990. CRI Publication No. 105-1990.

[69] National Aeronautics and Space Administration. Interior Landscape Plants for Indoor Air Pollution Abatement. September 15, 1989.

For Assistance and Additional Information

For assistance and guidance in dealing with known or suspected adverse effects of indoor air pollution, contact the U.S. Environmental Protection Agency Indoor Air Quality Information Clearinghouse (1-800-438-4318), EPA regional offices, and state and local departments of health and environmental quality, and your local American Lung Association (1-800-LUNG-USA).

For information on particular product hazards, contact the U.S. Consumer Product Safety Commission (1-800-638-CPSC). Individual manufacturers, as well as trade associations, may also supply pertinent information.

For information about regulation of specific pollutants, call the EPA Toxic Substances Control Act (TSCA) Assistance Information Service (202-554-1404).

For information relating to occupational exposures, contact the Occupational Safety and Health Administration (202-523-6091) or the National Institute of Occupational Safety and Health (1-800-35-NIOSH).

For information on lead, contact the National Lead Information Center (1-800-LEAD FYI). For information on pesticides, contact the National Pesticides Telecommunications Network (1-800-858-PEST).

Many sources of information are listed in the references at the end of each chapter. The following publications may also be useful to the health professional and to the patient.

General

For the health professional:

American Lung Association. "Health Effects and Sources of Indoor Air Pollution, Parts I and II". 1989. Publication No. 0857C.

American Thoracic Society. "Environmental Controls and Lung Disease". *American Review of Respiratory Disease.* 1990. 142: 915–939.

Gammage, R.B., Kaye, S.V. *Indoor Air and Human Health.* Lewis Publishers, Inc. Chelsea, MI.

Gergan, P.J., Weiss, K.B. "The Increasing Problem of Asthma in the United States". *American Review of Respiratory Disease.* 1992. 146(4): 823–824.

Gold, D.R. "Indoor Air Pollution". Clinics in Chest Medicine. June 1992. 13(2):215–229.

Samet, J.M., Spengler, J.D., eds. *Indoor Air Pollution — A Health Perspective.* Johns Hopkins University Press. Baltimore, MD. 1991.

Turiel, I. *Indoor Air Quality and Human Health.* 1985. Stanford University Press. Stanford, CA.

U.S. Environmental Protection Agency. "Building Air Quality: A Guide for Building Owners and Facility Managers". U.S. Government Printing Office. Washington, D.C. EPA-055-000-00390-4. December 1991.

U.S. Environmental Protection Agency. "EPA Indoor Environmental Quality Survey". 1992. OMB No. 2060-0244.

U.S. Environmental Protection Agency, U.S. Public Health Service, National Environmental Health Association. "Introduction to Indoor Air Quality: A Self-Paced Learning Module". EPA-400-3-91-002. July 1991.

U.S. Environmental Protection Agency. The Total Exposure Assessment Methodology (TEAM) Study; Project Summary. 1987. EPA-600-56-87-002.

Wadden, R.A., Scheff, P.A. *Indoor Air Pollution — Characterization, Prediction, and Control.* 1983. John Wiley and Sons, Inc. New York, NY.

For the patient (may be helpful to the professional as well):

American Lung Association. "Air Pollution In Your Home?". 1990. Publication No. 1001C.

American Lung Association. "Home Indoor Air Quality Checklist". 1992. Publication No. 0679C.

American Lung Association. "Indoor Air Pollution Fact Sheet — Household Products". 1990. Publication No. 1187C.

U.S. Environmental Protection Agency, U.S. Consumer Product Safety Commission. "The Inside Story: A Guide To Indoor Air Quality". 1993. EPA-402-R-93-013.

U.S. Environmental Protection Agency. "Targeting Indoor Air Pollution — EPA's Approach and Progress". September 1992. EPA-400-R-92-012.

Environmental Tobacco Smoke

For the health professional:

Bascom, R., Kulle T., Kagey-Sobotka A., Proud, D. "Upper Respiratory Tract Environmental Tobacco Smoke Sensitivity". *American Review of Respiratory Disease.* 1991. 143:1304–1311.

International Cancer Information Center. "Selected Abstracts on Environmental Tobacco Smoke and Cancer". National Cancer Institute. *Oncology Reviews* series. October 1989.

For the patient (may be helpful to the professional as well):

American Lung Association. "Indoor Air Pollution Fact Sheet - Secondhand Smoke". 1989. Publication No. 1185C.

American Lung Association. "Reducing the Health Risks of Secondhand Smoke". 1992. Publication No. 1085.

U.S. Environmental Protection Agency. "Respiratory Health Effects of Passive Smoking" Fact Sheet. January 1993.

U.S. Environmental Protection Agency. "What You Can Do About Secondhand Smoke". July 1993. EPA-404-F-93-004.

Combustion Products

For the patient (may be helpful to the professional as well):

American Lung Association. "Indoor Air Pollution Fact Sheet - Combustion Products". 1992. Publication No. 1182C.

U.S. Consumer Product Safety Commission, U.S. Environmental Protection Agency, American Lung Association. "What You Should Know About Combustion Appliances and Indoor Air Pollution". 1991. ALA Publication No. 3717C.

Carbon Monoxide

For the health professional:

Chaitman, B.R., Dahms, T.E., Byers, S., Carroll, L.W., Younis, L.T., Wiens, R.D. "Carbon Monoxide Exposure of Subjects With Documented Cardiac Arrhythmias". Health Effects Institute Research Report No. 52. 1992.

U.S. Consumer Product Safety Commission. "The Senseless Killer". 1993. GPO Publication No. 1993-0-356-764.

Kirkpatrick, J.N. "Occult Carbon Monoxide Poisoning". *Western Journal of Medicine.* 1987. 147:52–56.

Animal Dander, Molds, Dust Mites, Other Biologicals

For the health professional:

Burge, Harriet A. "Indoor Air and Infectious Disease". In: Cone, J.E., Hodgson, M.J. *Problem Buildings: Building-Associated Illness and the Sick Building Syndrome. State of the Art Reviews in Occupational Medicine.* 1989. 4(4): 713–721.

Burge, Harriet A. "Toxigenic Potential of Indoor Microbial Aerosols". In: Sandhu, S.S., MeMarini, D.M., Mass, M.J., Moore, M.M., Mumford, J.L. *Short-Term Bioassays in the Analysis of Complex Environmental Mixtures.* V. Plenum Publishing, Inc. 1987:391–721.

Gallup, J., Kozak, P., Cummins, L., Gillman, S. 1987. "Indoor Mold Spore Exposure: Characteristics of 127 Homes in Southern California with Endogenous Mold Problems". *Experientia Suppl.* 51: 139–142.

Health Department Victoria. 1989. "Guidelines for Control of Legionnaire's Disease". Health Department Victoria. Melbourne, Australia.

Morey, P.H., Feeley, J.C. Sr., Otten, J.A. "Biological Contaminants In Indoor Environments". STP 1071, Philadelphia: ASTM, 1990.

Platts-Mills, T.A.E., de Weck, A.L. 1989. "Dust Mite Allergens and Asthma — A Worldwide Problem (International Workshop)". *Journal Allergy Clinical Immunology.* 83: 416–427.

Pope, A.M., Patterson, R., Burge, Harriet A. "Indoor Allergens: Assessing and Controlling Adverse Health Effects. 1993. National Academy Press.

For the patient (may be helpful to the professional as well):

American Lung Association. "Indoor Air Pollution Fact Sheet - Biological Agents". 1991. Publication No. 1186C.

National Institutes of Health, National Institute of Allergy and Infectious Diseases. "Something in the Air: Airborne Allergens". March 1993. NIH Publication No. 93–493.

U.S. Consumer Product Safety Commission, American Lung Association,. "Biological Pollutants In Your Home". 1990. ALA Publication No. 3715C.

U.S Consumer Product Safety Commission, "News from CPSC - Portable Humidifiers Need Regular Cleaning During Winter Months". January 1992. Release No. 92–48.

U.S. Environmental Protection Agency. "Indoor Air Fact Sheet: Use and Care of Home Humidifiers". February 1991.

Tuberculosis

For the health professional:

"Diagnostic Standards and Classification of Tuberculosis". *American Review of Respiratory Disease.* 1990. 142:725–35.

"Prevention and Control of Tuberculosis Among Homeless Persons: Recommendations of the Advisory Committee for Elimination of Tuberculosis". Morbidity and Mortality Weekly Report. Centers for Disease Control and Prevention. 1992. 41 (No. RR-5): 13–23.

"Prevention and Control of Tuberculosis in Facilities Providing Long-Term Care to the Elderly: Recommendations of the Advisory Committee for Elimination of Tuberculosis. Morbidity and Mortality Weekly Report. Centers for Disease Control and Prevention. 1990. 39 (No. RR-10): 7–20.

"Prevention and Control of Tuberculosis in U.S. Communities with At-Risk Minority Populations: Recommendations of the Advisory Committee for Elimination of Tuberculosis". Morbidity and Mortality Weekly Report. Centers for Disease Control and Prevention. 1992. 41 (No. RR-5): 1–12.

"Screening for Tuberculosis and Tuberculosis Infection In High-Risk Populations: Recommendations of the Advisory Committee for the Elimination of Tuberculosis". Morbidity and Mortality Weekly Report. Centers for Disease Control and Prevention. 1990. 39 (No. RR-8): 1–7.

"The Use of Preventive Therapy for Tuberculosis Infection in the United States: Recommendations of the Advisory Committee for Elimination of Tuberculosis". Morbidity and Mortality Weekly Report. Centers for Disease Control and Prevention. 1990. 39 (No. RR-8): 8–12.

"Tuberculosis Among Foreign-Born Persons Entering the United States: Recommendations of the Advisory Committee for Elimination of Tuberculosis". Morbidity and Mortality Weekly Report. Centers for Disease Control and Prevention. 1990. 39 (No.R-18): 1–21.

"Tuberculosis and Human Immunodeficiency Virus Infection: Recommendations of the Advisory Committee for Elimination of Tuberculosis". Morbidity and Mortality Weekly Report. Centers for Disease Control and Prevention. 1989. 38(14): 236–238, 243-250.

For the patient (may be helpful to the professional as well):

American Lung Association. "Facts About The TB Skin Test". 1992. Publication No. 0178. (Spanish version, Publication No. 0177).

American Lung Association. "Facts About Tuberculosis". 1991. Publication No. 1091.

Volatile Organic Compounds

For the health professional:

Harving, H., Dahl, R., Molhave, L. "Lung Function and Bronchial Activity in Asthmatics During Exposure to Volatile Organic Compounds". *American Review of Respiratory Disease.* 143: 751–754.

Molhave, L., Bach, B., Pederson, O.F. 1986. "Human Reactions to Low Concentrations of Volatile Organic Compounds". *Environmental International.* 12:157–176.

Norback, D. et al. 1990. "Volatile Organic Compounds, Respirable Dust, and Personal Factors Related to the Prevalence and Incidence of the Sick Building Syndrome in Primary Schools". *Brit. J. Ind. Med.* 47:733–774.

Otto, D.A. et al. 1990. "Neurobehavioral and Sensory Irritant Effects of Controlled Exposure to a Complex Mixture of Volatile Organic Compounds". *Neurotox. and Texatol.* 12:1–4.

U.S. Environmental Protection Agency. Nonoccupational Pesticide Exposure Study (NOPES); Project Summary. Publication No. IAQ-0028.

Formaldehyde

For the patient (may be helpful to the professional as well):

American Lung Association. "Indoor Air Pollution Fact Sheet — Formaldehyde". 1989. Publication No. 1184C.

U.S. Consumer Product Safety Commission. "An Update On Formaldehyde". October 1990.

Sick Building Syndrome

For the health professional:

Berney, B.W., Light, E.N., Bennett, A.C. "Medical Evaluation of 'Building Related' Symptoms". Proceedings of the Seventh Annual Hazardous Materials Management Conference International. 1989.

Burge, S. et al. 1987. "Sick Building Syndrome: A Study of 4373 Office Workers". *Ann. Occupational Hygiene.* 31: 493–504.

Finnegan, M.J. et al. 1984. "The Sick Building Syndrome Prevalence Studies". *Brit. Medical Journal.* 289: 1573–1575.

Hedge, A. "Work-Related Illness In Offices: A Proposed Model of the Sick Building Syndrome". *Env. Int.* 15: 143–158.

Hodgson, V.S. et al. 1986. "The Sick Building Syndrome". In: Proceedings of the Third International Conference on Indoor Air Quality and Climate, Vol. 6. Evaluations and Conclusions for Health Sciences and Technology, pp. 87–97. Swedish Council for Building Research. Stockholm, Sweden.

Kreiss, K. 1989. "The Epidemiology of Building-Related Complaints and Illness". 572–592. In: *Problem Buildings: Building-Associated Illness and the Sick Building Syndrome.* Cone, J.E., Hodgson, M.E., eds. Hanley and Belfus, Inc., Philadelphia.

Mendell, M.J., Smith, A.H. "Consistent Pattern of Elevated Symptoms in Air-Conditioned Office Buildings: A Reanalysis of Epidemiologic Studies". *American Journal of Public Health.* 80: 1193–1199.

McCunney, R.J. "The Role of Building Construction and Ventilation in Indoor Air Pollution". *New York State Journal of Medicine.* April 1987. pp. 203–209.

Molina, C. et al. 1989. "Sick Building Syndrome — A Practical Guide". Report No. 4. Commission of the European Communities. Brussels, Luxembourg.

Norback, D. et al. 1990. "Indoor Air Quality and Personal Factors Related to the Sick Building Syndrome". *Scan. J. Work Environmental Health.* 16: 121–128.

Norback, D. et al. 1990. "Volatile Organic Compounds, Respirable Dust, and Personal Factors Related to the Prevalence and Incidence of the Sick Building Syndrome in Primary Schools". *Brit J Ind. Med..* 47: 733–774.

Seitz, T.A. 1989. "NIOSH Indoor Air Quality Investigations 1971–1988". 163-171. In: *The Practitioners Approach to Indoor Air Quality Investigations. Proc. Indoor Air Quality International Symposium.* Weedes, D.M., Gammage, R.B., eds. American Industrial Hygiene Association, Akron, Ohio.

Skov, P. et al. 1990. "Influence of Indoor Air Climate on the Sick Building Syndrome in an Office Environment". *Scandinavian Journal Work Environmental Health.* 16: 363–371.

Skov, P. et al. 1989. "Influence of Personal Characteristics, Job-Related Factors and Psychological Factors on the Sick Building Syndrome". *Scandinavian Journal Work Environmental Health.* 15: 286–295.

Skov, P. et al. 1987. "The Sick Building Syndrome in the Office Environment: The Danish Town Hall Study". *Env. Int.* 13: 334–349.

Stenberg, B. 1989. "Skin Complaints in Buildings with Indoor Climate Problems". *Env. Int.* 15: 81–84.

Sterling, T.D. et al. 1983. "Building Illness in the White Collar Workplace". *International Journal of Health Services.* 13:277–287.

U.S. Environmental Protection Agency. "Indoor Air Quality and Work Environment Study". EPA-21-M-3004. June 1991.

For the patient (may be helpful to the professional as well):

American Lung Association. "Indoor Air Pollution In The Office". 1992. Publication No. 1002C.

American Lung Association. "Office Indoor Air Quality Checklist". 1992. Publication No. 1003C.

U.S. Environmental Protection Agency. "Indoor Air Fact Sheet: Sick Building Syndrome". April 1991.

U.S. Environmental Protection Agency. "Indoor Air Fact Sheet: Ventilation and Air Quality In Offices". July 1990.

Asbestos

For the patient (may be helpful to the professional as well):

American Lung Association. "Indoor Air Pollution Fact Sheet - Asbestos". 1991. Publication No. 1188C.

Radon

For the health professional:

American Medical Association and U.S. Environmental Protection Agency. "Radon: The Health Threat with a Simple Solution. A Physician's Guide". AMA. EPA-402-K-93-008. 1993.

Fabrikant, J.I. "Shelter and Indoor Air in the Twenty-First Century — Radon, Smoking and Lung Cancer Risks". *Environmental Health Perspectives.* 1990. 86:275–280.

National Academy of Sciences. Comparative Dosimetry of Radon in Mines and Homes. National Academy Press. Washington, D.C. 1991.

National Research Council, Committee on the Biological Effects of Ionizing Radiation. "Health Risks of Radon and Other Internally Deposited Alpha-Emitters". *BIER IV.* Washington, DC: National Academy Press, 1988.

Nero, A.V., Jr. "Radon and Its Decay Products in Indoor Air: An Overview". In: Nazaroff, W.W., Nero, A.V., Jr., eds. *Radon and Its Decay Products In Indoor Air.* New York: John Wiley and Sons Inc. 1988: 1–53.

Roscoe, R.J. et al. "Lung Cancer Mortality Among Non-Smoking Uranium Miners Exposed to Radon Daughters". *Journal of the American Medical Association.* 262(5): 629–633. 1989.

Samet, J.M. "Radon and Lung Cancer". *JNCI.* 1989. 81: 745–757.

Samet, J.M., Stolwijk J., Rose, S.L. "Summary: International Workshop on Residential Radon Epidemiology". *Health Phys.* 1991. 60: 223–227.

U.S. Department of Health and Human Services, Public Health Service, Agency for Toxic Substances and Disease Registry, Radon Toxicity. 1992.

U.S. Environmental Protection Agency. National Residential Radon Survey: Summary Report. EPA-402-R-91-0111. 1992.

U.S. Environmental Protection Agency. Technical Support Document for the 1992 Citizens Guide to Radon. 1992.

For the patient (may be helpful to the professional as well):

American Lung Association. "Indoor Air Pollution Fact Sheet — Radon". 1992. Publication No. 1183C.

American Lung Association. "Facts About Radon: The Health Risk Indoors". Publication No. 0174C.

U.S. Environmental Protection Agency. "A Citizens Guide to Radon (second edition)". EPA-402-K-02-001. 1992.

Multiple Chemical Sensitivity

For the health professional:

Ashford, N.A., Miller C.S. 1991. *Chemical Exposures. Low Levels and High Stakes.* Van Nostrand Reinhold, New York.

Bell, Iris R. "Neuropsychiatric Aspects of Sensitivity to Low Level Chemicals: A Neural Sensitization Model". Conference on Low Level Exposure to Chemicals and Neurobiologic Sensitivity, Agency for Toxic Substances and Diseases Registry, Baltimore, MD, April 6-7, 1994. To be printed in *Journal of Toxicity and Public Health.*

Brooks, B.D. and Davis, W.F. 1991. *Understanding Indoor Air Quality.* CRC Press. Boca Raton.

Cullen, M.R. 1987. "The Worker with Multiple Chemical Sensitivities: An Overview", In: *Workers with Multiple Chemical Sensitivity, State of Art Rev. Occup. Med.* 2:669–681.

Heilman, B. "Multiple Chemical Sensitivity". *Chemical & Engineering News.* July 22, 1992.

Miller, Claudia S. "Chemical Sensitivity: History and Phenomenology". Conference on Low Level Exposure to Chemicals and Neurobiologic Sensitivity, Agency for Toxic Substances and Diseases Registry, Baltimore, MD, April 6–7, 1994. To be printed in *Journal of Toxicity and Public Health.*

Terr, A. "Clinical Ecology". *Annals of Internal Medicine.* 111(2): 168–178.

U.S. National Research Council. "Biologic Markers in Immunotoxicology". 1992. National Academy Press. Washington, DC.

U.S. National Research Council. "Multiple Chemical Sensitivities — Addendum to Biologic Markers in Immunotoxicology". 1992. National Academy Press. Washington, D.C.

Air Cleaners

For the patient (may be helpful to the professional as well):

U.S. Environmental Protection Agency. "Indoor Air Fact Sheet: Residential Air Cleaners". February 1990.

U.S. Environmental Protection Agency. "Residential Air Cleaning Devices: A Summary of Available Information". February 1990. Publication No. EPA-400-1-90-002.

Carpet

American Lung Association. "Indoor Air Pollution Fact Sheet — Carpet". 1992. Publication No. 1189.

U.S. Consumer Product Safety Commission. "Tips for Purchasing and Installing New Carpet" Fact Sheet. October 1992.

U.S. Environmental Protection Agency. "Carpet and Indoor Air Quality" Fact Sheet. October 1992.

☆ U.S. GOVERNMENT PRINTING OFFICE:1994-523-217/81322

Notes

Notes

Notes